## SUPERHERO ANIMALS
# FALCONS

**KENNY ABDO**

Fly!
An Imprint of Abdo Zoom
abdobooks.com

## abdobooks.com

Published by Abdo Zoom, a division of ABDO, P.O. Box 398166, Minneapolis, Minnesota 55439. Copyright © 2020 by Abdo Consulting Group, Inc. International copyrights reserved in all countries. No part of this book may be reproduced in any form without written permission from the publisher. Fly!™ is a trademark and logo of Abdo Zoom.

Printed in the United States of America, North Mankato, Minnesota.
102019
012020

Photo Credits: Alamy, Everett Collection, iStock, Library of Congress, Shutterstock, ©Copeinator123 Captain America Vol 1 117 p9 / CC-BY-SA
Production Contributors: Kenny Abdo, Jennie Forsberg, Grace Hansen
Design Contributors: Dorothy Toth, Neil Klinepier

### Library of Congress Control Number: 2019941307

### Publisher's Cataloging-in-Publication Data

Names: Abdo, Kenny, author.
Title: Falcons / by Kenny Abdo
Description: Minneapolis, Minnesota : Abdo Zoom, 2020 | Series: Superhero animals | Includes online resources and index.
Identifiers: ISBN 9781532129506 (lib. bdg.) | ISBN 9781098220488 (ebook) | ISBN 9781098220976 (Read-to-Me ebook)
Subjects: LCSH: Falcons--Juvenile literature. | Birds of prey--Juvenile literature. | Birds--Juvenile literature. | Raptors--Juvenile literature. | Zoology--Juvenile literature.
Classification: DDC 598.96--dc23

# TABLE OF CONTENTS

Falcons . . . . . . . . . . . . . . . . . . . . . . . . . 4

Origin Story . . . . . . . . . . . . . . . . . . . . . 8

Powers & Abilities . . . . . . . . . . . . . . 12

In Action . . . . . . . . . . . . . . . . . . . . . . . 18

Glossary . . . . . . . . . . . . . . . . . . . . . . 22

Online Resources . . . . . . . . . . . . . 23

Index . . . . . . . . . . . . . . . . . . . . . . . . . 24

Soaring from page to screen, Falcon has spread his wings and become one of the most beloved superheroes by fans.

As a fierce bird of **prey**, the falcon is known for its hunting skills. It is a ruthless **predator**.

Falcon swooped onto the scene in *Captain America* issue 117 in 1969. Stan Lee and artist Gene Colan came up with the character to be an **ally** of Captain America.

Lee and Colan also created Falcon in response to the **civil rights movement**. Doing so made him the first African-American to appear in a major comic book series.

# POWERS & ABILITIES

Falcons live all around the world. Falcons do not build nests. Instead, they live on cliffs or in trees. Some use nests built by other birds.

Falcons have long, pointed wings. This helps them make sharp turns in flight.

Peregrine falcons are the fastest animals on Earth. They can dive at speeds of more than 200 miles per hour (322 kmh)!

Falcons have large eyes. Their vision is estimated to be eight times better than a human's.

Falcons can spot **prey** from about two miles (3.21 km) away. They then swoop down accurately to catch their prey.

# IN ACTION

Sam Wilson is an ex-US Air Force **pararescue** airman. He is heavily trained in combat and marksmanship. Wilson is also a master pilot. He is also Falcon, the superhero.

Falcon uses a jet pack with attached wings called the EXO-7. This device allows him to fly at high speeds and with the **agility** of a falcon.

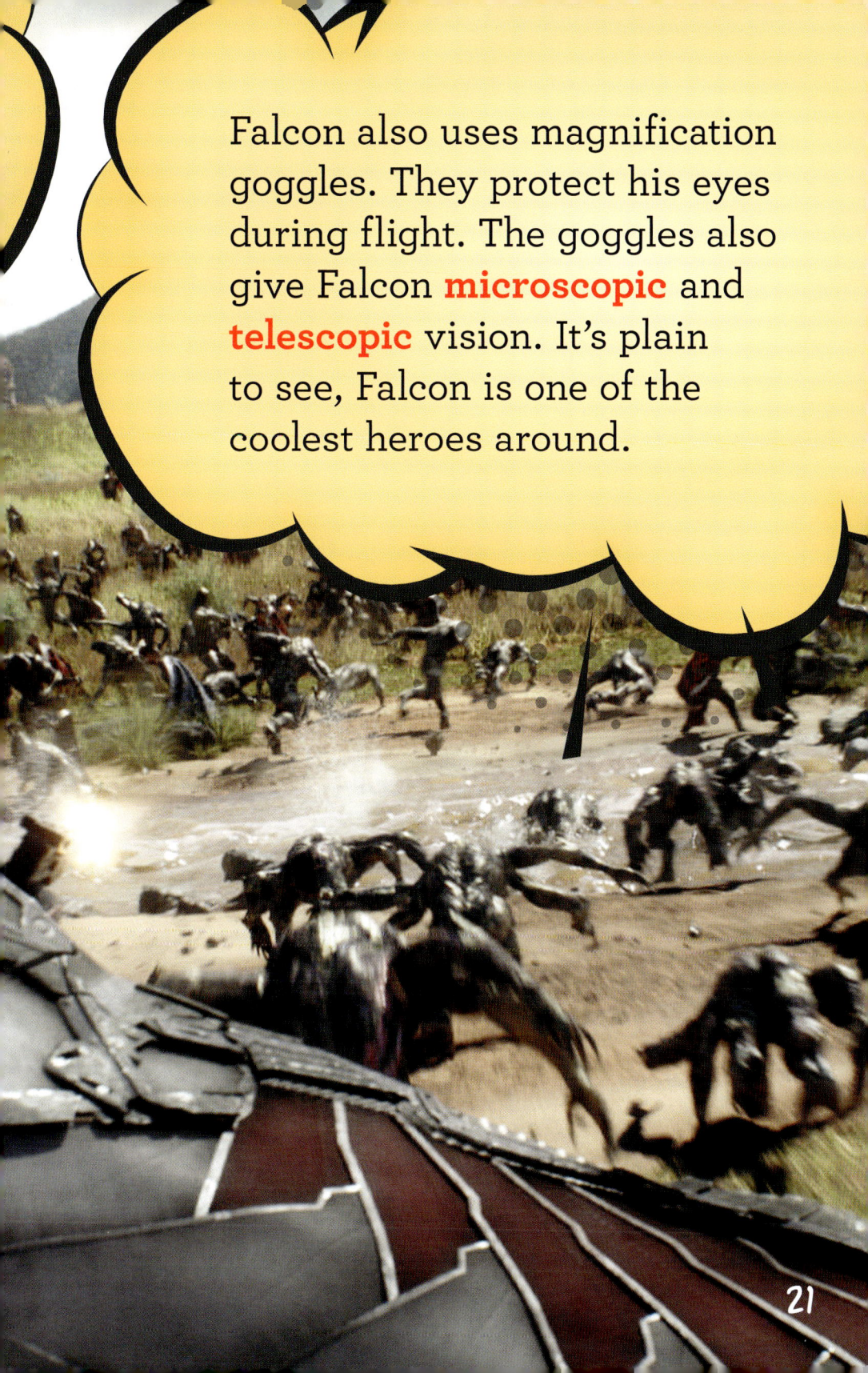

Falcon also uses magnification goggles. They protect his eyes during flight. The goggles also give Falcon **microscopic** and **telescopic** vision. It's plain to see, Falcon is one of the coolest heroes around.

# GLOSSARY

**agility** – the ability to move quickly and easily.

**ally** – a person, a group, or a nation united for some special purpose.

**civil rights movement** – a movement in the United States in the 1950s and 1960s. It consisted of organized efforts to end laws that involved unequal treatment of African Americans.

**microscopic** – something so small, it could only be seen through a microscope.

**pararescue** – Air Force specialists who rescue and medically treat hurt soldiers by parachuting, scuba diving, or rock climbing to them.

**predator** – an animal that lives by killing and eating other animals.

**prey** – animals hunted or killed by other animals for food.

**telescopic** – capable of seeing objects from an extreme distance.

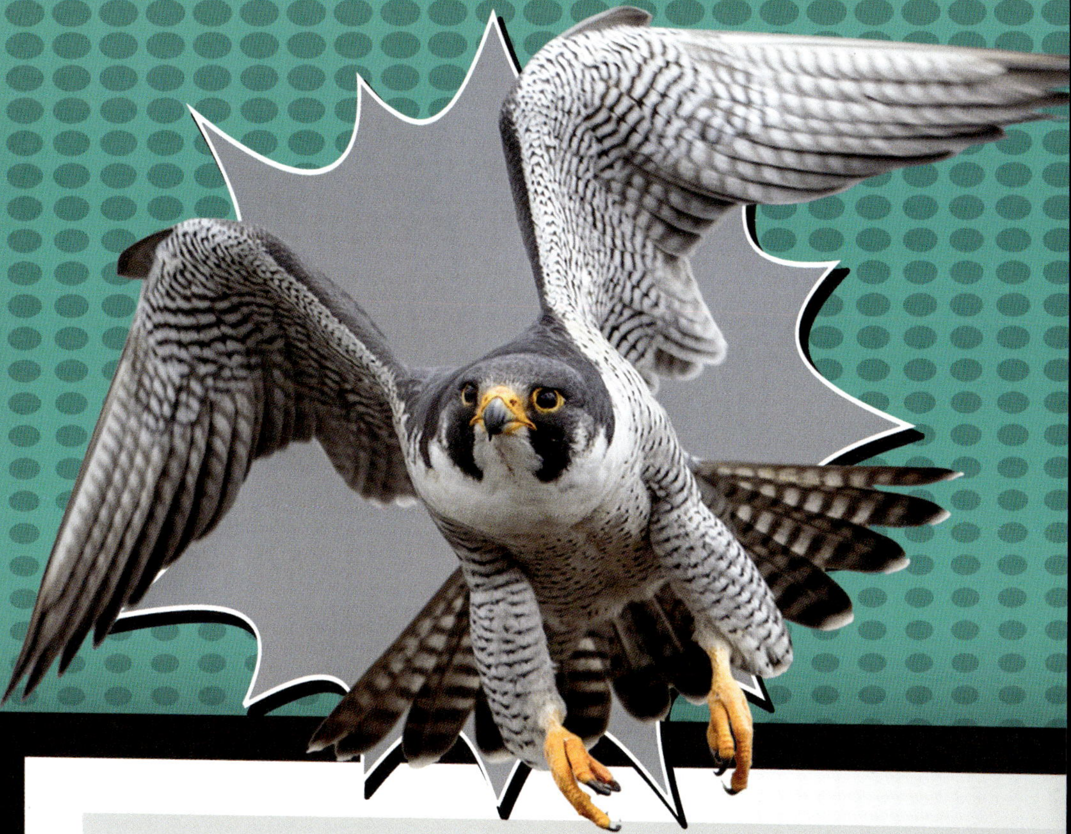

# ONLINE RESOURCES

To learn more about falcons, please visit **abdobooklinks.com** or scan this QR code. These links are routinely monitored and updated to provide the most current information available.

# INDEX

Captain America (character) 9

*Captain America* (comic) 9, 11

civil rights movement 11

Colan, Gene 9, 11

eyes 16, 17, 21

Falcon (character) 5, 9, 11, 18, 20, 21

flight 14, 17, 20

habitat 12

Lee, Stan 9, 11

US Air Force 18

Wilson, Sam (character) 18

wings 14, 20